A Collection of Poultry Books Owned by E.E. Richards

by E.E. Richards
of Cedar Rapids, Iowa

with an introduction by Jackson Chambers

Self Reliance Books

Get more historic titles on animal and stock breeding, gardening and old fashioned skills by visiting us at:

http://selfreliancebooks.blogspot.com/

Introduction

I am pleased to present yet another title on Poultry.

The work is in the Public Domain and is re-printed here in accordance with Federal Laws.

As with all reprinted books of this age that are intended to perfectly reproduce the original edition, considerable pains and effort had to be undertaken to correct fading and sometimes outright damage to existing proofs of this title. At times, this task is quite monumental, requiring an almost total "rebuilding" of some pages from digital proofs of multiple copies. Despite this, imperfections still sometimes exist in the final proof and may detract from the visual appearance of the text.

I hope you enjoy reading this book as much as I enjoyed making it available to readers again.

Jackson Chambers

Early Agriculture, Natural History and Ornithology, All of Which Have Parts Devoted to Poultry

WILLUGHBEII (FRANCISCI). Ornithologiae, libri tres; in quibus aves Omnes hactebus Cognitae in Metgodum Naturis suis Convenien tim redactae Aceurate Describuntur, Descriptiones Iconbus, Ele gantissimis, etc., Totum opus Recognovit, Digessit Supplevit Joannes Raius with vignette, pp. 307 and 77 plates folio, original panelled calf, Londini 1671. The plates illustrate over 300 species, with both English and Latin names on each.

J. W. (John Wolbridge). Gent. Systema Agriculturae; The Mystery of Husbandry discovered. Treating of the several new and most advantageous ways of tilling, planting, sowing, manuring, ordering, improving, of all sorts of gardens, orchards, meadows, pastures, corn lands, woods and coppices, as also of fruits, corn, grain, pulse, new hays, cattle, fowl, beasts, bees, silk worms, fish, etc. Also containing dictionarium rustivum, or the interpretation of rustic terms. London, 1687. Leather, pp. 326.

VANIERII (JACOBI). Praedium Rusticum. Editio Novissima. Coloniae Munatianae 1750. 12mo, half vellum, pp. 320 and index. Many wood cuts.

ALDROVANDI (ULYSSIS). Philosophi et Medici Bononiensis Ornithologiae. Bononiae 1600, folio, orig. calf, engraved title and port., many wood cuts of Poultry.

DER ARBEITSAME LAND und Haussvater, order Kurtser Anterricht, wie ein Land und Haussvater. Regenspurg, 1710, 16mo, pp. 234.

LIBRI DE RE RUSTICA. By Catonis, Varronis, Columellae, Palladii, Enarratio and Petri. Lugduni Apud Seb. Gryphium, 1541, thick 12mo, half calf. Woodcuts.

REAUMURS (HERRN DE). Anweisung wie man zu jeder Jahreszeit allerlen zahmes Gefiugel entweder vermittelst der Barme des Mistes oder des gemeinen Feuers ausbruten und aufziehen solle. Trans. from the French by M. Johann Christoph Thenn. Augsburg, 1767, 12mo, wra., pp. 272. Folding plates.

BEWICK (THOMAS). A General History of Birds and Quadrupeds. With 128 engravings. Phila., 1824. Bd. in calf, pp. 142.

CUVIER (BARON GEORGUS, Councillor of France and Minister of Public Instruction). The Animal Kingdom. Arranged after its organization, forming a natural history of animals and an introduction to comparative anatomy. Translated and adapted to the present state of science. With additions by W. B. Carpenter, M. D., F. R. S. and J. O. Westwood, F. L. S. Illustrated by three hundred engravings in wood and thirty-four on steel. London, 1851. Half leather, pp. 717.

BUFFON (GEORGE LUOIS, Count of Buffon). A Natural History, general and particular; containing the History and Theory of the Earth, A General History of Man, The Brute Creation, Vegetables, Minerals, etc. Translated from the French by William Smellie. New edition corrected and enlarged to which is added A History of Birds, Fishes, Reptiles, and Insects, embracing the recent discoveries of eminent naturalists together with an account of the most curious foreign plants by Henry Augustus Chambers, LL. D. In two vols. Containing page plate G. L. LeClerc, Count De Buffon. Colored illustrations. Plates of the fish in 2nd vol, all hand colored. London, (n. d.), cloth, I vol. pp. 682, II vol. pp. 662.

COLUMELLA (L. JUNIUS MODEBATUS). Of Husbandry in Twelve Books and his Book concerning Trees. Translated into English, with several illustrations from Pliny, Cato, Varro, Palladius, and other ancient and modern authors. London, MDCCXIV. Calf, pp. 600.

COOKE (George). The Complete English Farmer; or, Husbandry made perfectly easy, in all its useful branches, containing what every farmer ought to know and practice. Among the various articles treated of in this work is the following, viz.: The whole are of Rearing and Managing Fowls, Ducks, Geese, Turkeys and Pigeons to make them turn out profitable to the Farmer, with choice receipts to cure their several distempers. London, (n. d.). Half morocco, pp. 165.

COBBETT (WILLIAM). Cottage Economy. Containing information relative to the brewing of Beer, making the Bread, keeping of Cows, Pigs, Bees, Ewes, Goats, Poultry, etc., also instructions for erecting and using ice houses, after the Virginian manner. Fifteenth edition. London, 1838. Orig. bds., 1838, pp. 199.

CHOYSELAT (M. PRUDENT LE.) Discovrs Oeconomique, non moins utile que recreatif, monstrant comme de cinq cens liures pour une foys employees, l'on peult tirer par an quatre mil cinq cens liures de proffict honneste, qui est le moyen de faire prosier son argent. Paris, 1612. Narrow 12mo, calf, pp. 45.

GAYOT (Eug.). Poules et Oeufs. Paris, n. d. 12mo, orig. wra., pp. 208, illus., many on steele.

HERREA (GAB. ALFONSO D'). Agricoltvra Pratta da Diversi Antichi et Moderni Scrittori. Et Tradotta di Lingva Spagnuola in Ilatiana. Da Mambrino Roseo da Fabriano. Venetia, 1592. Sq. 8vo, orig. velleum.

JARDINE (SIR WILLIAM, F. B. S. E., F. L. S., etc., etc.). The Naturalist's Library. Ornithology, Vol. III. Gallinaceous Birds. Beautiful colored illustrations of Gallenaceous Birds, especially of the Meleagris O ceilata. Also, 112 pages of Memoir of Aristotle. Edinburgh, W. H. Lizars and Stirling and Kenney, Longman, Rees, Orme, Browne, Green and Longman. London, and W. Curry Jun & Co., Dublin, 1834. Orig. bd., pp. 232.

MARKHAM (GREVESE). Cheap and Good Husbandry. For the well ordering of all beasts and fowls, and for their general cure of their diseases. 1683. 156 pages. Calf. Printed by T. B. for Hannah Sawbridge, at the Bible on Ludgate-Hill, London.

PLINIVS (CLAUDIUS, Secvdndvs). The Historie of the World. Commonly called, the Natvrall Historie of C. Plinivs Secvndvs. Translated into English by Philemon Holland Doctor of Physicke. The first Tome. London, 1635. Printed by Adam Islip, and are to be fold by John Grismond, in Ivy-Lane at the Signe of the Gun. Half leather, pp. 632.

2

REAUMUR (M. DE). **Pratique de L'art de Faire Eclorre** et d'elever en toute saison des Oiseaux Domestiques de toutes especies, etc. Paris, 1751. Narrow 12mo, calf, pp. 144, folding plates.

REES (ABRAHAM, D. D., F. R. S., F. L. S. Amer. So.). **The Cyclopedia or Universal Dictionary of Arts, Sciences, and Literature.** Illustrated with numerous engravings by the most distinguished artists. First American Edition. Revised, corrected, enlarged and adapted to this country by several literary and scientific characters. Plates Vol. 5. Natural History. Phila. Orig. bds., pp. not given.

RAY (JOHN, late fellow of the Royal Society). **The Wisdom of God Manifested in the Works of the Creation.** In two parts, viz.: The Heavenly Bodies, Elements, Meteors, Fossils, Vegetables, Animals, (Beasts, Birds, Fishes and Insects), more particularly in the Body of the Earth, its Figure, Motion, and Consistency, and in the admirable Structure of the Bodies of Man, and other Animals, as also in their Generation, etc., with answers to some objections. Containing full page steele plate of Joannes Rajus. The sixth edition, corrected. London, printed for William Innys, at the Prince's Arms in St. Paul's Church Yard, 1714. Bound in calf, pp. 405.

SMELLIE (WILLIAM, member of Antiquarian and Royal Societies of Edinburgh). **The Philosophy of Natural History.** Containing book plate Ormathwaite. London, MDCCXC (1790). Calf, pp. 547.

TUSSER (THOS.). **Five Hundred Points on Husbandrie**, newly set foorth by Thomas Tusser, Gentleman. 1586. 168 pp., calf. At London, printed in the now dwelling house of Henrie Denham in Aldergate Street at the Sign of the Starre. Contains book plate of Henry Francis Lyte.

LYDEKKER (RICHARD). **Library of Natural History,** with introduction by Ernest Thompson-Seton. 72 full page color plates and 2,000 engravings. 1904. 6 vols., cloth, 3356 pp.

Books Dealing With Poultry Exclusively

ABBOTT (H. E.). **Monthly Hints for Indian Fowl Fanciers.** Calcutta, India, 1908. Paper covers, pp. 91.

ALLEN (R. L.). **Domestic Animals,** History and Description of the Horse, Mule, Cattle, Sheep, Swine, Poultry, and Farm Dogs. With directions for their management, breeding, crossing, rearing, feeding and preparation for a profitable market, also their diseases and remedies, together with full direction for the management of the Dairy. Illustrated. New York, 1849. Limp cloth, pp. 227.

AMERICAN POULTRY JOURNAL. **Origin and History of all Breeds of Poultry.** Trustworthy information regarding the origin and history of all recognized varieties of chickens, ducks and geese. 80 pages, boards, full page color plates, 1908. American Poultry Journal, Chicago, Ill.

ARBUTHNOTT (THE HON. MRS.). **The Henwife.** Her experience in her own Poultry Yard. With illustrations by Harrison Weir. Seventh edition, enlarged. Edinburgh, 1868. Limp cloth, pp. 220.

ASSOCIATED FANCIERS. **Fanciers' Hand-Book (No. 3).** The Practical Poultry Book, for both the Farmer and Fancier. Containing illustrations and autograph H. S. Babcock. Phila., (n. d.). Paper covers, pp. 100.

BY AN INDIANA HENWIFE. **My poultry and How I Manage Them.** Third edition. Calcutta, 1900. Paper covers, pp. 55.

ATKINSON (HERBERT). The Old English Game Fowl. Its History, Description, Management, Breeding and Feeding. Edition De Luxe. Illustrations in black and white. London, (n. d.). Cloth, pp. 65.

AYRES (F. H.). Quest of the Leghorn. A Book of Few Theories, Many Facts. Containing illustration by Porter. Hartford, Conn., 1880. Paper covers, pp. 40.

AYRES (F. H.). The Plymouth Rock as the Fowl for General Use. With rules for mating and breeding, according to nature. Containing illustration. Hartford, 1878. Paper covers, pp. 28.

AYRES (F. H.). The Game Fowl, Its Origin and History. With Rules for Mating, Rearing and Training for the Exhibition Room of the Pit. Containing illustrations. Hartford, 1878. Paper covers, pp. 30.

AYRES (F. H.). The Leghorn of the Past and Present. Containing illustrations. Hartford, 1878. Paper covers, pp. 28.

AUTHOR NOT GIVEN. Poultry for Profit. Being a complete guide to the profitable management of Domestic Fowls including Turkeys, Geese and Ducks, shoinwg how a large income may be obtained therefrom. With colored and other illustrations. With autograph "Caroline J. A. B. Mortimer, Seghill House, November 25, 1911. London, (n. d.). Orig. bds., pp. 180.

W. H. V. How to Raise Poultry on a Large Scale. Showing plans of buildings, layouts of runs, methods of feeding and taking necessary care of fowls on a poultry farm. Containing autograph of H. S. Babcock. Hartford, 1883. Paper covers, pp. 64.

AMERICAN BREEDERS' ASSOCIATION, Vol. V. Report of the meeting held at Columbia, Mo., January 6, 7, and 8, 1909, and for the year ending December 31, 1908. Wash., D. C., (n. d.). Pp. 420.

A PHONOGRAPHIC REPORT of the Meeting of Breeders and Experts, held in Boston, March 7-14, 1885. Boston, 1885. Paper cover, pp. 125.

G. P. The Lady Housekeeper's Poultry Yard. Its pleasure and profit. Containing a clipping pasted in by Mr. Gaunt. London, (n. d.). Limp cloth, pp. 91.

AUTHOR OF A POULTRY FARMER. Poultry for Exhibition, Home and Market. With a chapter on Pheasants and Pheasantries. Illustrated. Frontispiece, "Japanese Bantam." London, (n. d.). Cloth, pp. 95.

AUTHOR OF BRITISH HUSBANDRY. Farming for Ladies; or, a guide to the poultry yard. the Dairy and Piggery. Illustrated. London, 1844. Cloth, pp. 511.

AUTHOR NOT GIVEN. Supposed to be Ridgeway. A Treatise on the Breeding, Rearing and Fattening of Poultry. Second edition. Containing book mark G. B. K. Also autograph on fly leaf. J. B. Kingdon. Dedicated to Lord Somerville by the editor. London, 1819. Original paper covers rebound, pp. 196.

AUTHOR NOT GIVEN. Supposed to be Ridgeway. A Treatise on the Breeding, Rearing and Fattening of Poultry. Second edition. Dedicated to Lord Somerville. Containing book mark, G. A. Dincley Goodyere. London, 1819. Bound in calf, pp. 196.

AUTHOR NOT GIVEN. Profits in Poultry. Useful and ornamental breeds and their profitable management. Profusely illustrated. New York, 1908. Cloth, pp. 352.

3

AUTHOR NOT GIVEN. Supposed to be Ridgeway. A Treatise on the Breeding, Rearing and Fattening of Poultry. Chiefly translated from the New French Dictionary on Natural History. London, 1810. Half leatrer, pp. 196.

AUTHOR NOT GIVEN. Profits in Poultry. Useful and ornamental Breeds and their Profitable Management. Profusely illustrated. New York, 1886. Cloth, pp. 256.

AUTHOR NOT GIVEN. The Raising and Management of Poultry. With a view to establishing the best breeds; the qualities of each as egg and flesh producers; their care and profit; and the great and increasing value of the poultry interests to farmers and the country.

AUTHOR NOT GIVEN. The Poultry Doctor. Including the homeopathic treatment and care of chickens, turkeys, geese, ducks and singing birds; also a materia medica of the chief remedies. Phila., 1891. Cloth, pp. 85.

AUTHOR NOT GIVEN. Eggs and Poultry as a Source of Wealth. Illustrated. Containing autograph of T. Purcell, 1866. London, (about 1866). Paper covers, pp. 91.

AUTHOR NOT GIVEN. Book of Household Pets, and How to Manage Them. Containing complete and valuable instructions about the Diseases, Breeding, Training and Management of the Canary, Mocking Bird, Brown Thrush, etc., and all kinds of Pigeons and Poultry, etc. Illustrated with 123 fine woodcuts. New York, (n. d.). Orig. bds., pp. 116.

AUTHOR NOT GIVEN. Cows in India and Poultry; Their Care and Management. Second edition. Illustrated. Calcutta, 1896. Orig. bds., pp. 159.

AUTHOR NOT GIVEN. The Pleasures and Profits of our Little Poultry Farm. London, 1879. Cloth, pp. 95.

A PRACTICAL POULTERER. The New England Poultry Breeder. Being a brief history of Domestic Fowls, and containing full directions for their rearing and management. Illustrated with twenty-five correct engravings. The practical Poulterer was Geo. P. Burnham, who states in his "Hen Fever," that this book was written, illustrated, type set and printed complete in one week. Boston, 1850. Orig. bds., pp. 107.

AN ASSOCIATION OF PRACTICAL BREEDS. The American Fowl Breeder; containing full information on Breeding, Rearing, Diseases and Management of Domestic Poultry; also, instructions concerning the choice of pure stock, crossing, caponizing, etc., with engravings. With appendix of 3 pages containing report of the first American Poultry show. Boston, 1850. Orig. bds., pp. 88.

AUTHOR NOT GIVEN. Eggs all the Year Round at Four Pence per Dozen. And Chickens at Four Pence per Pound. Containing full and complete information for the successful and profitable keeping of poultry. Third edition. London, 1878. Paper covers rebound, orig. bds., pp. 96.

BABCOCK (H. S.). A Poultry Compendium. Being a brief treatise on the Rearing and Management of Domestic Fowls. Contains on the back cover illustration Pea-Comb Barred Plymouth Rock cockerel, Pilgrim VI. Owned by H. S. Babcock, Providence, R. I. Hartford, 1885. Paper covers, pp. 62.

BABCOCK (H. S.). The Indian Game. Its Description, Characteristics, Origin, History and Breeding. Containing colored illustration by Lee. Fort Wayne, 1891. Paper covers, pp. 25.

BABCOCK (H. S.). The Rhode Island Red Fowl. From the annual report, 1899, of the Rhode Island State Board of Agriculture. Providence, 1900. Paper covers, pp. 16.

BABCOCK (H. S.). The Argonaut, the Ideal General Purpose Fowl. A brief treatise upon the Breed. Danielsonville, Conn., 1891. Paper covers, pp. 24.

BACON (G. W., F. R. G. S.). Guide to Success in Poultry Keeping, showing how to make poultry pay in summer and winter; with many new and valuable hints, and 200 golden rules. Containing illustrations. London, (n. d.). Orig. bds., pp. 122.

BARTON (F. F.). Every Day Ailments of Poultry. The aim of the present little work is to supply the amateur and professional poultry keeper with the most modern, correct and practical information regarding the "every-day ailments, and their treatment," in poultry. Illustrated. Perth, (n. d.). Cloth, pp. 80.

BASLEY (A.). Mrs. Basley's Poultry Book. *Western Poultry Book* Tells you what to do and how to do it. The chicken business from first to last including 1,001 questions and answers. Los Angeles, Calif. Paper covers, pp. 192.

BATES (MORGAN). Why Poultry Pays, and How to Make it Pay. 48 pages, illustrated, boards, 1889. American Poultry Journal, Chicago, Ill.

BAUM (L. FRANK). The Book of the Hamburgs. A brief treatise upon the Mating, Rearing and Management of the different varieties of Hamburgs. Containing autograph of H. S. Babcock. Hartford, 1886. Paper covers, pp. 71.

BAILY (JOHN). The Dorking Fowl. Containing hints for its management and feeding for the table. London, (1853). Limp cloth, pp. 39.

BAILY (JNO.). Fowls. A plain and familiar treatise on the principal breeds, with reprint of the fourth edition of Dorking fowl (second edition). 1854. 86 pages, limp covers. Henningham & Hollis, London, Eng.

BAILY (JOHN). Fowls. A plain and familiar treatise on the principal breeds. Instructions for breeding and exhibition. Fifth edition, revised, corrected and enlarged. With which is reprinted, The Dorking Fowl; Its Management and Feeding for the Table. Seventh edition. Autograph of Helen S. Clark. London, 1868. Bound in limp cloth, gilt edged, pp. 151.

BAILY (JNO.). Fowls. 1875. Sixth edition. 164 pp. Bds. Henningham & Hollis, London, Eng.

BAILEY (JOHN). Fowls. A plain and familiar treatise on the principal breeds. Instructions for breeding and exhibition. Sixth edition, revised, corrected and enlarged. With which is reprinted The Dorking Fowl—Its Management and Feeding for the Table. Eighth edition. London, 1876. Cloth, bds., pp. 164.

BANNAN (B.). The Poultry Breeder's Text Book. Comprising full information respecting the choicest breeds of poultry and the mode of rearing them. With twenty-five illustrations. Pottsdam, Pa., about 1850. Paper covers, pp. 76.

BEALE (STEPHEN). Profitable Poultry Keeping. Edited with additions by Mason C. Weld, who was first secretary of N. Y. S. P. A., with original illustrations. With clipping pasted in on the Weight and Yield of Eggs. New York, (n. d.). Cloth bds., pp. 258.

BEMENT (C. N.). The American Poulterer's Companion. A practical treatise on the Breeding, Rearing, Fattening and general management of the various species of Domestic Poultry, with illustrations and portraits of fowls taken from life. New York. 1845. Limp cloth, pp. 379. Autograph of Wm. Fitch, 1845. (First edition).

——— **SAME**, 1852-1863.

BENNETT (JNO. C., M. D.). **The Poultry Book.** A treatise on Breeding and general management of Domestic Fowls, with numerous original descriptions and portraits from life.

BOSTON: Phillips, Sampson & Company, New York; Chas. M. Saxton, Philadelphia; Thomas Cowperthwait & Co., Baltimore; Cushing & Brother, Charleston; S. C. McCarter & Allen, Cincinnati; H. W. Derby & Co., Buffalo; G. H. Derby & Co., 1850. Limp cloth, pp. 310. (First edition). Appendix containing report of first poultry show held in America, report of the organization, election of officers, constitution and by-laws of the first poultry organization in America, viz., The New England Society for the Improvement of Domestic Poultry.

BEUOY (GEORGE). **What's a Capon and Why.** Containing illustrations. Cedar Vale, Kas., 1911. Paper covers, pp. 40.

BEETON (S. O.). **Beeton's Book of Poultry and Domestic Animals,** showing how to rear and manage them in sickness and in health. Colored frontispiece by H. Wier. London, about 1861. Cloth, gilt edged, pp. 332.

BEEVER (REV. W. HOLT, M. A.). **The Daily Life of Our Farm.** Containing frontispiece by S. C. Swain, "The Pets of the Household." London, 1871. Leather and cloth bds., pp. 313.

BIGGS (FRED). **Minorca Fowls.** Their Breeding and Management for Eggs and for the Show Pen. With the Standard of Perfection of the London Minorca Club, and of the American Black Minorca Club. Fifth edition, revised and enlarged. With colored plate and numerous engravings. London, (n. d.). Paper covers, pp. 45.

BIGGLE (JACOB). **Biggle Poultry Book.** A concise and practical treatise on the management of farm poultry. Illustrated. Philadelphia, 1909. Cloth cover, pp. 162.

BICKEL (M. V.). **Poultry Packers Guide.** A compendium of useful information for poultry dressers. M. V. Bickel, Mason City, Ia.

BISHOP (A. V.). **Poultry Keeping.** An ideal book on poultry keeping for utility or exhibition. 1901. 50 pp., boards, illustrated. Mark & Moody (Ltd.), Stowbridge, Eng.

BJERGAARD (J. PEDERSEN. **Dansk-Fjerkrae-Standard** (Danish Standard). Illustrated. 1908. Limp cloth, pp. 264.

BJERGAARD (J. PEDERSEN). **Vort Fjerkrae.** Illustrated. 1901. Cloth, pp. 178.

BLAIR (MRS. FERGUSSON). **The Henwife.** Her Own Experience in her own poultry yard. With colored illus. by Harrison Weir. Containing book mark "R. L. L." Edinburgh, 1861. Limp cloth, gilt edges, pp. 192.

BLAIR (MRS. FERGUSSON). **The Henwife.** Her Own Experience in her own poultry yard. With illustrations by Harrison Weir. Third edition. Containing photo grauver of Mrs. Arbuthnot. Edinburgh, Thomas C. Jack, 82 Princess St. London, Hamilton, Adams & Co., 1862. Cloth, pp. 218.

BLAIR (J. GAYLOR). **Poultry Diseases and Their Remedies.** The cause, symptoms and treatment of all diseases known to poultry. 100 pp., 1909, boards. J. Gaylor Blair, Lexington, Ky.

BOSWELL (PETER). **The Poultry Yard.** A practical view of the method of selecting, rearing and breeding the various species of domestic fowl. New edition. London, MDCCXLV. Cloth, pp. 200.

BOYEE (MICHAEL K.). Profitable Poultry Farming. 1905. 47 pp., paper covers.

BOYER (MICHAEL K.). A Living from Poultry. A treatise written from actual experience, showing how to make start in poultry farming with limited capital. 1910. 40 pp., paper covers.

BOYER (MICHAEL K.). All About Broilers and Market Poultry Generally. 1902. Ills., 39 pp., paper covers.

BOYER (MICHAEL K.). Money in Hens. A small treatise showing how hens and pullets can be kept with a profit. 1895. 34 pp., paper covers.

BOYER (MICHAEL K.). Money in Broilers and Squabs. Together with special chapters on turkey and guinea broilers and green ducklings and geese for market. 1904. 148 pp., ills., paper covers.

BRANFORD (EDGAR). The Malay Fowl and Malay Bantam. With notes on the fancy in Australia. By Tom Cadwell. Containing illustrations. London, 1894. Paper covers, pp. 55.

BRIGHAM (ARTHUR A.). Sharpe (S. C.). Progressive Poultry Culture. The keeping of poultry for profit and pleasure. With twenty-four illustrations. London, (n. d.). Limp cloth, pp. 227.

BRIGHAM (A. A.). Chick Book. A carefully prepared work covering the subject of chick raising. 80 pp., boards, illustrated. 1908. Hawkins Pub. Co., Waterville, N. Y.

BRIGHAM (DR. ARTHUR A., B. S., Ph. D.). Progressive Poultry Culture. A text book of study and practice in the keeping of poultry for profit and pleasure. Originally appeared in a serial article of twelve chapters in Western Poultry Journal. Cedar Rapids, Iowa. Cloth, pp. 292.

BROWNE (D. J.). The American Poultry Yard. Comprising the origin, history and description of the different breeds of domestic poultry with complete directions for their breeding, crossing, rearing, fattening and preparation for market; including specific directions for caponizing fowls, and for the treatment of the principal diseases to which they are subject; drawn from authentic sources and personal observation. Illustrated with numerous new engravings. With an appendix embracing the comparative merits of different breeds of fowls. By Samuel Allen. New York, 1860. Cloth, pp. 333.

BROWNE (D. J.). The American Poultry Yard. Comprising the origin, history and description of the different breeds of domestic poultry with complete directions for their breeding, crossing, rearing, fattening and preparation for market; including specific directions for caponizing fowls and for the treatment of the principal diseases to which they are subject. Drawn from authentic sources and personal observation. Illustrated with numerous engravings. With an appendix, embracing the comparative merits of different breeds of fowls by Samuel Allen. Book plate of Wm. B. Denning. New York, 1850. Cloth, pp. 332. (First edition).

BROWN (EDWARD). F. L. S. Poultry-Keeping, as an industry for farmers and cottagers. Sixth edition. Illustrated. Frontispiece by Beynon. London, 1906. Cloth, pp. 206).

BROWN (EDWARD). F. L. S. Poultry. Their Varieties, Classification, Exhibiting, Treatment, Breeding, Rearing, Housing, Diseases, and General Management. With twenty-four illustrations by Ludlow and frontispiece showing their points. Containing autograph of H. S. Babcock. London, (n. d.). Paper board covers, pp. 136.

5

BROWN (EDWARD). F. L. S. Poultry Keeping, as an industry for farmers and cottagers. Illustrated by Ludlow. London, 1891. Cloth, pp. 138.

BROWN (EDWARD). F. L. S. Wright (Francis H.). F. S. A. A. National Poultry Conference held at Reading on July 11th, 12th, and 13th, 1899. Office report. London, 1899. Cloth, pp. 138.

BROWN (EDWARD). F. L. S. Second National Poultry Conference held at University College, Reading, July 8, 9, 10 and 11, 1807. Official report. Illustrated. London, 1907. Cloth, pp. 382.

BROWN (EDWARD). F. L. S. Races of Domestic Poultry. 234 pages, cloth, fully illustrated. 1906. Edward Arnold. London, Eng.

BROWN (EDWARD). The Poultry Industry in Denmark and Sweden. A complete report of methods. 1908. 112 pages, illustrated. National Poultry Organization Society, London, Eng.

BROWN (EDWARD). Report of Poultry Industry of America. 1907. 124 pp., boards. National Poultry Organization Society, London. Eng.

BROWN (ROSE, M. C. A.). Poultry and Game. A handbook of useful and practical recipes for choosing, preparing, cooking and serving all kinds of poultry and game. London, (n. d.). Paper covers, pp. 29.

BROWN (J. T., F. Z. S.). (Chanticleer). Encyclopedia of Poultry. All that its name infers. 536 pp., cloth, bound in two vols. 1910. Walter Southwood & Co. (Ltd.), London.

BROOMHEAD (WILLIAM W.). Poultry and Profit. With eight full-page plates. London, 1911. Cloth, pp. 128.

BROOMHEAD (WM. W.). The Poultry Club Standards. Containing a complete description of all the recognized varieties of fowls, ducks, geese and turkeys. Fourth edition. Cassell, A. & Co., Ltd. London, New York, Toronto and Melbourne, 1910.

BURN (ROBERT SCOTT). Outlines of Modern Farming. Vol. IV. The Dairy—Pigs—Poultry. With notes on the diseases of stock. Eighth edition. London, 1901. Cloth, pp. 211.

BURNHAM (GEO. P.). The China Fowl. Shanghae, Cochin and "Brahma." With forty choice illustrations. Contains the original photograph and autograph of Geo. P. Burnham. Melrose, Mass., 1874. Cloth, pp. 168.

BURNHAM (GEO. P.). The History of the Hen Fever. A humorous record in one volume, illustrated. Dedication: To the Amateurs, Fanciers and Breeders of Poultry, the successful and unfortunate dealers, throughout the United States; and the Victims of misplaced confidence in the Hen Trade, generally, I Dedicate This Volume. Boston, James French & Co.; New York, J. C. Derby; Philadelphia, T. B. Peterson (N. D.). Cloth, pp. 326.

BURNHAM (GEO. P.). Burnham's New Poultry Book. A practical work on selecting, housing and breeding domestic fowls. Illustrated with drawings of modern popular varieties, plans of poultry houses, etc. Boston, (n. d.). Cloth, pp. 342.

CANTELO (WM. JAS.). Cantelonian System of Hatching Eggs. Hydro incubation or top contact heat. 1849. 40 pp., original paper cover, unbound in boards. Wm. Strange, London, Eng.

4

"CHANTICLEER". Poultry. Illustrated. London, 1908. 12mo, cloth, pp. 124.

CHRISTY (J., Jr.). Hydro-Incubation. By means of which all kinds of poultry and game birds may be inexpensively hatched and successfully reared all the year round. Illustrated. London, 1877. Paper covers, pp. 19.

CLARKE (H. P.). A. M., M. D. Rules of the Cock Pit. Containing illustrations. Indianapolis, 1900. Paper covers rebound with bds., pp. 56.

CLEWETTE (FRANK B.). Poultry West of the Rockies. This book is the experience in a condensed form of hundreds of poultrymen in the west and is designed to point out the obstacles to poultry raising and now to overcome them. Los Angeles, 1902. Bds., pp. 123.

CLOUGH (W. W.). Plymouth Rocks. How to Mate and Breed Them. Solid advice by our leading breeders. Containing illustrations. Danielson, Conn., (n. d.). Paper covers, pp. 38.

CLOUGH (W. W.). Clough's Bantam Book. How to Mate, Breed and Care for them. Devoted exclusively to bantams. colored and uncolored illustrations. Medway, (n. d.). Paper covers, pp. 64.

COATES (ROBERT). How to Succeed with Small Fruits and Poultry. With illustrations. Showing the advantage of combining these two pursuits with instructions for raising and management of both. Also diseases and their remedies. Chicago, 1885. Paper covers, pp. 110.

COBURN (F. D., Secretary of Kansas State Board of Agriculture). Profitable Poultry. Report of the Kansas State Board of Agriculture for the quarter ending September, 1908. Devoted to descriptions and illustrations of the land and water fowls, most generally reared in America, with directions for their breeding, maintenance and profitable management. Contains five anatomical charts. Topeka, 1900. Original cover rebound in cloth, pp. 322.

COBB (E., F. Z. S.). Poultry Farming Up To Date. Strand, London, W. C., (n. d.). Paper covers, pp. 32.

COBB (E., F. Z. S.). Breeding Poultry for Exhibition. Strand, London, W. C. Paper covers, pp. 136.

COCK (MICAJAH R.). The American Poultry Book. Being a practical treatise on the management of domestic poultry. New York, 1843. Cloth, pp. 180.

COMYNS (Alexander, B. A., LL. B.). Scientific Breeding and Feeding. (Originally published as "Fancy Poultry"). Second edition. London. Paper covers, pp. 48.

COMYNS (ALEXANDER, B. A., LL. B.). Hon. Sec. Poultry Club, Editor of "Poultry," etc. Issued by the authority of the Poultry Club. The Standard of Perfection for Exhibition Poultry. Part I. Illustrated. Autograph of H. S. Babcock. London, W. C., (n. d.). Paper covers, pp. 62.

CONOVER (M. ROBERTS). Making a Poultry House. Illustrated. New York, 1912. Cloth, pp. 54.

CORBETT (PROF. A.). The Poultry Yard and Market. A practical treatise on Gallinoculture, and description of a new process for hatching eggs and raising poultry. Containing photo and autograph of the author. New York, 1877. Paper covers rebound with bds., pp. 96.

CORBIN (F. H.). Plymouth Rocks. Their Origin, Characteristics, Requirements, etc. With special reference to the improved strain. Containing photo and autograph of the author; also autograph of J. B. Lowe, Fortville, Ind. Hartford, 1879. Paper covers, pp. 94.

COOK (P.). Successful Incubation. A working manual for hatching plants. 1911. 36 pages, boards, illustrated. Weimer Press, Los Angeles, Cal.

COOPER (J. W., M. D.). Game Fowls. Their origin and history with description of breeds, strains and crosses. American and English modes of feeding, training and heeling. 1869. 304 pages, cloth, illustrated in colors. J. W. Cooper, West Chester, Pa. Contains book plate of J. S. Dudley, Lynchburg, Va.

COOK (WM.). Fowls for the Times. The history and developments of the Orpington folw. Published by the author at Orpington House, St. Mary Cray, Kent, London Agent; E. W. Allen, 4 Ave. Maria Lane, London E. C. (n. d. Cloth, pp. 165.

COOK (WM.). Practical Poultry Breeder and Feeder; or, How to Make Poultry Pay. Published by the author at Orpington House, St. Mary Cray, Kent, England, and Scotch Plains, New Jersey, U. S. A., (n. d.). Cloth, pp. 270.

CROAD (A. C.). The Langshan Fowl: Its History and Characteristics with some comments on its early opponents. (3rd edition). Containing illustrations. London, 1889. Cloth, pp. 122.

CROAD (A. C.). The Langshan Controversy, in England and the United States. England, 1879. Paper covers, pp. 28.

CRAIG (FRANCIS D.). The Complete Poultry Manual. Illustrated. North Evanston, Ill., (n. d.). Paper covers, pp. 57.

CURTIS (GRANT M.). Cyphers Series. Capons, Profitable Marketing of Poultry, Profitable Egg Farming, Profitable Poultry Houses and Appliances, Profitable Care and Management, Profitable Poultry Keeping. Six bound as one, 756 pages, illustrated, buckram. 1900. Cyphers Incubator Co., Buffalo, N. Y.

CYPHERS (CHAS. A.). Eggs, Broilers and Roasters. An easy lesson in practical poultry culture. 1906. 64 pages, boards, illustrated. Chas. A. Cyphers, Buffalo, N. Y.

CYPHERS (CHARLES A.). Incubation and its Natural Laws. 1891. Paper covers rebound, orig. bds., pp. 111.

DAVENPORT (C. B.). Inheritance of Poultry. A publication of the Carnegie institution of Washington, illustrated with 17 pages of plates. 1906. 134 pages, boards. Carnegie Institution, Washington, D. C.

DAVENPORT (GEO. M.). Practical Experience with Poultry. New York, (n. d.). Paper covers, pp. 32.

DAVIS (J. H.). The Art of Poultry Breeding. A discussion of out-crossing, in-breeding, breeding to feather, and cross-breeding for market purpose. Chatham, N. Y., (n. d.). Paper covers, pp. 46.

DAVIS (J. H.). The Possum Creek Poultry Club. Illustrated. Chatham, N. Y., 1895. Paper covers, pp. 109.

DAVIES (C. J.). Poultry. A Handbook for Poultry Keepers. Illustrated. London, 1907. Orig. bds., pp. 120.

DAVISON (B. W.). Practical Poultry Culture. A concise practical treatise on the management of poultry for profit. Indianapolis, 1898. Paper covers, pp. 144.

DE REAUMUR (M.). The Art of Hatching and Bringing Up of Domestic Fowls of all Kinds at any Time of the Year. Either by means of the heat of hot-beds, or that of common fire. Illustrated. London, MCCCL. Rebound in calf, pp. 47.

DE LANCY (FRANK W.). A to Z of Pigeons and Bantams. 1910. 97 pages, bound, illustrated. Item Pub. Co., Sellersville, Pa.

DICKSON (WALTER B.). Poultry. Their Breeding, Rearing, Diseases and General Management. Frontispiece Illustration of Bankiva Cock; Polish cock and hen; Spanish cock and hen. London, 1838. Cloth, pp. 316.

DICKSON (WALTER B.). Mrs. London, Poultry; their Breeding, Rearing, Diseases and General Management. New edition, incorporating the treatise of Bonington Moubray. With illustrations by Harvey. Frontispiece in color. London, (n. d.). Cloth, pp. 262.

DIXON (Rev. E. S.). Ornamental and Domestic Poultry; their History and Management. Thick post 8vo, 1848. Presentation copy to J. Gould, the eminent ornithologist with author's inscription on fly-leaf.

DICKIE (A. M., M. D.). Diseases of Poultry; How to Avoid and Cure Them. Con. autograph H. S. Babcock. New York, 1880. Paper covers, pp. 42.

DIEHL (JOHN E.). The Poultry Doctor. A treatise on diseases of poultry with symptoms and remedies, homeopathic and Allopathic; also a chapter on keeping and rearing poultry. Phila., 1879. Paper cover rebound, orig. bds., pp. 34.

DIXON (REV. E. S.). The Dovecote and the Aviary. Sketches of the Natural History of Pigeons and other Domestic Birds in a Captive state. With hints for their management. Containing illustrations. London, (n. d.). Cloth, pp. 458.

DOYLE (MARTIN). The Illustrated Book of Domestic Poultry. The figures drawn from nature by C. H. Weigall; engraved and printed in oil colors by W. A. Dickes & Co. Phila., (n. d.). Cloth, pp. 382.

DOYLE (MARTIN). The Illustrated Book of Domestic Poultry. The figures drawn from nature by C. H. Weigall; engraved and printed in oil colors by W. A. Diches & Co. London, 1854. Limp cloth, pp. 114.

EDWARDS (KINARD B.). How the French Make Fowls Pay. In four parts, bound in one. 1871. 44 pages, orig. paper covers rebound boards, illustrated. Thos. Bosowrth, London, Eng.

ELLETT (A. E.). Modern Wyandottes. How to breed, Manage and exhibit. 62 pages. N. d. Boards, illustrated black and colors. Poultry Press (Ltd.), London, Eng.

ELKINGTON (W. M.). Egg and Poultry Raising at Home. Illustrated. London, 1902. Rebound orig. bds., pp. 92.

ELKINGTON (W. M.). Popular Poultry Keeping for Amateurs. A practical and complete guide to breeding and keeping poultry for eggs or the table. 1907. 140 pages, paper, ills. L. Upcott Gill, London, Eng.

ENGLEFIELD (H.). New Laid Eggs all the Year Round. from our own Garden and House Scraps. Or How to Manage and feed Fowls in Confined Runs, for Eggs. With autograph of Rachel Ellis. Crandon, (n. d.). Paper covers, rebound bds., pp. 31.

ENTWISLE (WILLIAM FLAMANK). Bantams. Illustrated by Ludlow. Wakefield, (n. d.). Cloth, pp. 132.

EVANS (ERNEST). The Biology of Poultry Keeping; or the Domestic Fowl, its History, Anatomy, Food, Reproduction and Breeding. With illustrations. Frontispiece, Gallus Bankica, the Indian Jungle cock, drawn by W. E. Holt. from a specimen in the Natural History Museum, South Kensington. London, 1899. Cloth, pp. 108.

Morris-Eyle

EYLE (MORRIS, L. C. R.). Brahmas and Cochins. Illustrated. With a preface by Lewis Wright. London, 1899. Paper covers, pp. 73.

FANCIERS' REVIEW. Five Hundred Questions and Answers on Poultry Raising. A book of practical and authentic information in the form of questions and answers on various subjects, as feed and care, diseases, eggs, incubators, buildings, etc., with a chapter on turkeys, geese and ducks. Second edition. Containing autograph of H. S. Babcock, 324 Binter Exchange, Prov., R. I. Chatham, 1892. Paper covers, pp. 60.

FARM JOURNAL SERIES. Contains:$6.41 per Hen. Poultry Secrets, Curtis Poultry Book, Duck Dollars, Turkey Secrets, Million Dollar Egg Farm, six bound as one. 470 pp., orig. cover rebound in buckram, illustrated. 1909-11. Farm Journal, Philadelphia, Pa.

FELCH (I. K.), Babcock (H. S.)., Lee (Henry J.). The Philosophy of Judging. A manual upon the scoring of exhibition fowls; intended to meet the wants of the general breeders, and the exhibitor as well as the professional judge. Fort Wayne, 1889. Cloth, pp. 217. (Containing autograph of H. S. Babcock).

FELCH (I. K.). The Amateur's Manual; or Specific Mating of Thoroughbred Fowls. Containing autograph of H. S. Babcock and autograph of the author. Boston, 1877. Orig. bds., pp. 110.

FELCH (I. K.). Poultry Culture. How to Raise, Manage, Mate and Judge Thoroughbred Fowls. Illustrated. Containing photograph and autograph of I. K. Felch. Chicago, 1886. Cloth, pp. 430.

FELCH (I. K.). Standard American Perfection Poultry Book, describing all the different varieties of fowls, their points of beauty and their merits as setters. Chicago, 1903. Paper covers rebound in orig. bds., pp. 164.

FELCH (I. K.). The Breeding and Management of Poultry; or Thoroughbred for Practical Use. Presentation copy to E. E. Richards by the author, I. K. Felch. Hyde Park, 1877. Pp. 88, orig. bds.

FESSENDEN (THOS. G.). Moubray's a Treatise on Breeding, Rearing and Fattening all kinds of Poultry, Cows, Swine and other Domestic Animals. Reprinted from the sixth London edition with such abridgements and additions, as was conceived would render it best adapted to the soil, climate and common course of culture in the United States by Thomas G. Fessenden. Boston, published by Lilly & Wait, and Carter & Hendee. 1832. Frontispiece illustration from Bewick. First poultry book published in America. Cloth, pp. 266.

FESSENDEN (THOMAS G.). Moubray on Breeding, Rearing and Fattening all Kinds of Poultry, Cows, Swine and other Domestic Animals. Second American, from the sixth London edition. Adapted to the Soil, Climate and Culture of the United States. Frontispiece illustration by Bewick. Boston, Joseph Breck & Co., New York; G. C. Thornburn, 1837. Cloth, pp. 278.

FERGUSON (G.). Illustrated Series of Rare and Prize Poultry. Including comprehensive essays upon all classes of Domestic Fowl. The figures drawn and colored from prize specimens by C. J. Culliford. Dedicated to the Earl of Derby by George Ferguson. London, 1854. Limp cloth, pp. 372.

FERRIS (J. F.). Practical Artificial Incubation. A resume of the progress made in the past few years in artificial incubation in this country and Europe with descriptions of a score of leading incubators in successful operation in America and England. Also chapters upon the proper care and management of the young chicks. Profusely illustrated. Albany, 1880. Limp cloth, pp. 110.

FIELD (J. PENFOLD). The Wyandotte Fowl. Containing autograph to Harmon S. Babcock with the author's Compliments and regards Fbry. 1891. With illustrations. London, (n. d.). Paper covers, pp. 53.

FIELD (FANNY). Poultry for Market and Poultry for Profit. Twelve articles. Written expressly for those who are interested in poultry and wish to make it profitable. Chicago, 1885. Paper covers, pp. 47.

FINN (FRANK, B. A.), (Oxon), F. Z. S., M. B. O. U. The Water Fowl of India and Asia. Illustrated. Calcutta, 1909. Papers covers, pp. 121.

FINCHLEY MANUALS OF INDUSTRY, No. V. Domestic Fowls and Animals. Their Natural History, Rules for their Breeding and Rearing; with the Management of the Dairy. Directions for Keeping Bees, etc. Prepared for the use of the National and Industrial Schools of the Holy Trinity, at Finchley, Middlesex. London, MDCCLIV. Cloth, pp. 135.

FINN (FRANK) B. A. (Oxon) F. Z. S., M. B. O. U. How to Know the Indian Waders. Calcutta, 1906. Papers covers, pp. 223.

COLLETT (J. W.). The A. B. C. Guide to Rearing Poultry in India. Containing valuable information on everything appertaining to poultry-keeping in India. Bombay, (n. d.). Paper covers, pp. 126.

FRASER'S MAGAZINE. Containing Clippings from Tegetmeir's Profitable Poultry. Poultry Pentalogue, and Alectryomania. London, Eng., 1854. Paper covers.

GAMMERDINGER (CHAS.). Der Amerikanische Huhnerhof. A book of practical poultry culture in German. 1878. 40 pp., boards, illustrated. Chas. Gammerdinger, Columbus, O.

GEYELIN (GEO. KENNEDY, C. E.). Geyelin's Poultry Breeding. In a commercial point of view. As carried out by the National Poultry Company (Ltd.), Bromley, Kent. Natural and Artificial Hatching, Rearing and Fattening. On entirely new and scientific principles with all the necessary plans, elevations, sections and details and a notice of the poultry establishments in France. With a preface by Charles L. Flint. With twenty-seven illustrations. First edition. N. Y., 1867. Limp cloth, pp. 127.

GRAY (D. F. Thompson). ("Psyche"). Poultry Ailments and Their Treatment. For the use of Amateurs, illustrated. Dundee, James P. Mathew & Co., 17 Cowgate, 1885. Orig. bds., pp. 56.

GRAY (F. H.). Cocker's Manual. Devoted to the Game Fowl. Their origin and breeding, rules for feeding, heeling, handling, etc. Description of the different breeds, diseases and their treatment. Second edition (revised). Illustrated. Battle Creek, Mich., 1878. Cloth, pp. 155.

HAIG (J. B.). Common Sense in the Poultry Yard. A Story of failures and success, including full account of 1,000 hens and what they did. 1885. 192 pp., cloth, illustrated. Industrial Pub. Co., New York.

HARE (FRANK C.). Built and used by Poultrymen. A book describing and illustrating practical houses and appliances. 1909. 94 pp., boards, illustrations. The Standard Co., Quincy, Ill.

8

HAWKS (EARL B.). Science and Art of Poultry Culture. A practical text-book of poultry husbandry in its various branches. Illustrated. Clinton, Wis., 1909. Cloth, pp. 490.

HECK (FRANK). 999 Questions and Answers. A guide to success with poultry, written and arranged in the form most helpful. 1903. 126 pp., boards. Frank Heck, Chicago, Ill.

HASTINGS (MILO M.). The Dollar Hen. Illustrated. Containing three pages of press comments. New York, 1909. Cloth, pp. 217.

HECK (FRANK). Secrets of Expert Exhibitors and Easy Lessons in Judging. An exposition of the methods employed by breeders of Standard Bred Fowls in preparing their birds for poultry shows, including many dishonest schemes which are occasionally practiced. Containing signed pledge No. 605, not to loan or reveal any of the secrets contained therein. Chicago, 1909. Morocco, pp. 78.

HEWES (THEO.). How to Make Poultry Pay. Money making information covering the experience of many recognized authorities. 48 pages, boards, illustrated in colors. 1909. Inland Poultry Journal Co., Indianapolis, Ind.

HEWES (THEO.). With Historical Notes by Dr. H. P. Clark. Hamburgs, Wyandottes, Rhode Island Reds. From shell to show room. Four bound in one. 192 pp., boards, illustrated. Black and colors. 1905. Inland Poultry Journal, Indianapolis, Ind.

HICKS (J. STEPHEN). Ewart (H. G.). The Possibilities of Modern Poultry Farming. Being a review of the poultry industry. A description of the popular breeds and notes on feeding, as originally appearing in the columns of "Farm Life." London, (n. d.). Paper covers, pp. 82.

HICKS (REESE V.). Tricks of the Poultry Trade. Some methods, little things and "nigh cuts" practiced among the "initiated" of the craft. 1909. 64 pp., boards, illustrated. Copper Pub. Co., Topeka, Kan.

HOME COUNTIES. Poultry Farming. Some facts and some conclusions. Illustrated. Frontispiece (from "The Country Gentleman and Land and Water"). A champion layer. The White Wyandotte which in 1902-3 competition produced 78 eggs in 16 (winter) weeks. New York, 1905. Cloth, pp. 186.

HOWE (JOHN B.). India Runner Duck Culture from "A to Z." Containing illustrations. Fortville, Ind. Papers covers, pp. 32.

HOWARD (GEORGE E.). The American Fancier's Poultry Book. With illustrations by the author. Wash., D. C., (n. d.). Cloth, pp. 169.

HURST (J. W.). Successful Incubation and Brooding. A parctical guide to the hatching and rearing of poultry by artificial means. London, (n. d.). Flexible cloth, pp. 113.

HURST (J. W.). The Life Story of a Fowl. An annual autobiography. 1908. 220 pages. Illustrated in colors. Cloth. Adam & Chas. Black, London.

HYDE (D. D.). Poultry and Eggs for Market and Export. Published by direction of Hon. T. J. Duncan, minister of agriculture. 1905. 82 pp., boards, illustrated. John Mackey, Gov. Ptr., Wellington, N. Z.

JACKSON (LAWRENCE). Poultry for Profit. Illustrated. 1st edition. Haysville, Pa., 1910. Paper covers rebound in orig. bds., pp. 84.

JAMES (MRS. ELIOT). Profitable and Economical Poultry-Keeping. Illustrated. London, (n. d.). Cloth, pp. 140.

JAMES (ED.). The Game Cock. Being a practical treatise on Breeding, Rearing, Training, Feeding, Trimming, Mains, Heeling, Spurs, etc., together with an exposure of Cocjers' Tricks. The origin and cure of diseases, and the revised cocking rules governing all parts of the world. Illustrated. N. Y., 1873. Cloth, pp. 71.

JOHNSON (A. T.). Chickens and how to Raise Them. All about chickens, how to hatch, house, feed and fatten them and cure their diseases. Philadelphia, 1910. Cloth, pp. 159.

JOHNSON (Mrs. Rebecca). How to Hatch, Brood, Feed and Prevent Chicks from Dying in the Shell. Revised second edition. Maxwell, Ia., 1906. Cloth, pp. 64.

JOHNSON (PROF. WILLIS GRANT). Brown (George O.). The Poultry Book. By many expert American breeders and Harrison Weir, F. R. H. S. Illustrated from drawings in color and black and white. By Weir and from photographs. In three volumes. New York, 1904. Cloth, three vols., pp. 1299.

JOHNSON (G. M. T.). Practical Poultry Keeping As I Understand It. Fourth edition. Illustrated. Containing autograph "H. S. Babcock." Binghampton, N. Y., 1884. Paper covers, pp. 96.

JOHNSTONE (E. B.). The A. B. C. of Poultry. A reference work for Amateur, Fancier and Professional on Poultry Keeping. Containing a catalogue of general literature published by Sir Isaac Pitman & Sons, Ltd. London, 1906. Cloth, pp. 166.

KAINS (M. G.). Profitable Poultry Production. Illustrated. New York, (n. d.). 1910. Pp. 278.

KERR & DIXON (REV. EDMUND SAUL, A. M., Rector of Intwood, with Kehwick, Norfolk). A Treatise on the History and Management of Oriental and Domestic Poultry. With large additions by J. J. Kerr, M. D. Illustrated with 65 original potralts engraced expressly for this work. Philadelphia, 1855. Limp cloth, pp. 480.

KELLERSTRASS (ERNEST). The Kellerstrass Way. (First edition). 1910. 94 pp., illustrated. Ernest Kellerstrass, Kansas City, Mo.

KELLERSTRASS (ERNEST). The Kellerstrass Way of Raising Poultry. 234 pages, boards, illustrated. 1902. Author also publisher.

LEE (J. HENRY). Some of Lee's Ideas. Practical hints for those who would help themselves in the construction of conveniences for use about the yard, the garden and the farm, with especial reference to poultry-keeping. Illustrated. Indianapolis, 1894. Paper covers, pp. 93.

LEE (CHARLES). A Practical Guide for the Breeding, Feeding, Rearing and General Management for Domestic Use and Exhibition of the Houdan fowl. Illustrated. Containing a plate of the author's Houdan cock "Lionel" from a photograph.

LEWIS (WM. M.). The People's Practical Poultry Book. A Work on the Breeds, Breeding, Rearing and General Management of Poultry. Illustrated with over one hundred engravings. Containing an appendix, 13 pages. New York, 1871. Limp cloth, pp. 189.

LILLIE (FRANK R.). The Development of the Chick. An introduction to Embryology. Illustrated. New York, 1908. Cloth, pp. 472.

LONG (JAMES). Poultry for Prizes and Profit. Being practical details for the Breeding, Management and Exhibition of Domestic Fowls. Illustrated. London, 1886. Cloth, pp. 204.

LONG (JAMES.) Elkington (W. M.). Poultry for Prizes and Profit. A complete and practical guide to the breeding and management of all varieties of poultry for exhibition and utility purposes. Illustrated. London, 1909. Cloth, pp. 188.

LONG (JAMES). Elkington (W. M.). Poultry for Prizes. A complete and practical guide to the breeding and management of all varieties of poultry for exhibition. Being Division 1 of poultry for prizes and profit. Illustrated. London, 1909. Cloth, pp. 191.

LOUDON (MRS.). Poultry, their Breeding, Rearing, Diseases and General Managemsnt. With illustrations by Harvey. Containing autographs of W. G. Cole, Newbold Verdon. London, 1853. Cloth, pp. 262.

LUPTON (F. M.). The Standard American Poultry Book. A Guide to Profitable Poultry Keeping. New York, 1886. Paper covers rebound in bds., pp. 128.

LUSHINGTON (J. I.). The Poultry Yard. How to "Farm" it to make the "Crop" pay. London, 1866. 48 pp., boards. (Under title "Chickens").

MARTIN (W. C. L.). Our Domestic Fowls. Containing on fly leaf Wm. Chas. Linnalus Main, son of John Martin, Supt. of Zoological Society, London, 1830 to 1838. Born 1798, died 1864. This book published 1847. London, 1847. Cloth, sm. 12mo, pp. 192.

MAIN (JAMES, A. L. A., P. A. S.). Domestic Poultry. A Treatise on the Breeding, Rearing and Fattening of Poultry, with additions and wood cuts. Fourth edition. Four pages of announcements of works on agriculture. London, MDCCL. Cloth, pp. 340.

McFetridge (G. A.). Poultry. A concise treatise on all branches. How to Hatch, Feed, Brood and Prepare for Market. Syracuse, 1897. Paper covers, pp. 80.

McDOUGALL (F. W.). Treatise on the Game Cock on Breeding, Rearing, Training, Feeding, Trimming, Heeling, Handling Diseases and their Treatment, Gaffs, etc., together with rules of the Pit. Containing illustration. Indianapolis, 1879. Paper covers, pp. 53.

McGREW (T. F.). Ths Egg Question Solved. Illustrated. Washington, D. C., 1904. Cloth, pp. 27.

McGREW (T. F.). The Feather's Plymouth Rock Book. Containing colored and uncolored illustrations. Wash., D. C., (n. d.). Pp. 70.

McGREW (T. F.). The Feather's Wyandotte Book. Containing colored and uncolored illustrations. Wash., D. C., (n. d.). Cloth, pp. 63.

McGREW (T. F.). The Perfected Poultry of America. A Concise, illustrated treatise of the recognized breeds of poultry, turkeys, and water fowl. Wash., D. C., 1907. Limp cloth, pp. 247.

McGREW (T. F.). Ths Bantam Fowl. Description of all standard breeds and varieties of bantams and of new breeds that are becoming popular. Fully illustrated. 68 pp., boards. Reliable Poultry Journal, Quincy, Ill. 1903.

McGREW (T. F.). How to Grow Chicks. Illustrated. Wash., D. C., (n. d.).

MINER (T. B.). Miner's Domestic Poultry Book. A treatise on the history, breeding and general management of foreign, and domestic fowls, embracing all the late importations of fowls and being descriptions by the best fowl fanciers in the United States, of all the most valuable breed with the author's extensive experience as breeder, together with selected matter of interest, comprising, as it is believed, the most complete and authentic work on the subject ever published. Illustrated by numerous portraits from life. Rochester, N. Y. Published by Geo. W. Fisher, also A. S. Barnes & Co., New York; B. B. Mussey, Boston; J. W. Moore, Philadelphia; J. B. Steele, New Orleans; H. W. Derby, Cincinnati. 1853. Forward by Jno. C. Bennett. Limp cloth, pp. 256.

MITTEL (J. B.). The Schemer. A sure way to prevent young chicks from dying in the shell. Big Springs, Tex. ,(n. d.). Paper covers, pp. 16 .

MOFFATT (J. M.). The Poultry-Keeper's Guide. Practical methods of Breeding, Rearing and Feeding all kinds of Poultry, including Cotchin China and other Fowls; and of preventing or curing the diseases to which they are subject. Seventh edition enlarged and improved from information furnished by a celebrated poultry breeder. London, about 1850. Orig. bds., pp. 52.

MOORE (JNO.). Columbarium; or The Pigeon-House. Being an introduction to a Natural History of Tame Pigeons, giving an account of the several species known in England, with the method of breeding them, their distempers and cures. London, printed for J. Wilford, behind the Chapter-House in St. Paul's Church-Yard, 1735. Reprinted by Jos. M. Wade, Fanciers' Journal Office. Philadelphia, 1874. Limp cloth, pp. 64.

MORANT (MAJOR G. F.). How to Keep Laying Hens and to Rear Chickens, in large or small numbers, in absolute confinement, with "Perfect Success." London, (n. d.). Paper covers, pp. 32.

MYRICK (HERBERT). Turkeys and How to Grow Them. A treatise on the natural history and origin of the name of turkeys; the various breeds and best methods to insure success in the business of turkey growing. With essays from practical turkey growers in different parts of U. S. and Canada. Copiously illustrated. New York, 1897. Cloth, pp. 159.

MOUBRAY (BONINGTON). A Treatise on Domestic Poultry, Pigeons and Rabbits. With a practical account of the Egyptian method of hatching eggs by artificial heat; and all the needful particulars relative to breeding, rearing and management. With agricultural reports of Great Britain. Dedicated to her grace the Duchess Dowager of Rutland. First edition. Lon., 1815. 12mo, orig. bds., pp. 218.

MOUBRAY (BONINGTON). A Practical Treatise on Breeding, Rearing and Fattening all Kinds of Domestic Poultry, Pheasants, Pigeons and Rabbits. With an account of the Egyptian method of hatching eggs, by artificial heat. Second edition, with additions, on the breeding, feeding and management of swine, from memorandum made during forty years' practice. Containing 10 pages of description of books published by Sherwood, Neely & Jones. London, 1816. 12mo, orig. bds., pp. 256.

MOUBRAY (BONINGTON). A Practical Treatise on Breeding, Rearing and Fattening all Kinds of Domestic Poultry, Pheasants, Pigeons and Rabbits. Including an interesting account of the Egyptian method of hatching eggs by artificial heat; with some modern experiments thereon. Containing autograph of Jno. C. Barrett, London 28-d Nov. 1820. Third edition. Lon., 1819. 12mo, orig. bds., pp. 288.

MOUBRAY (BONINGTON). A Practical Treatise on Breeding, Rearing and Fattening all Kinds of Domestic Poultry, Pheasants, Pigeons and Rabbits. Fourth edition. Lon., 1822. 12mo, morocco, solid gilt edges, sides and back, colored illustrations, orig. bds., pp. 312.

MOUBRAY (BONINGTON). A Practical Treatise on Breeding, Rearing and Fattening all Kinds of Domestic Poultry, Pheasants, Pigeons and Rabbits. Colored illustrations. With the announcement of a number of early publications on agriculture. Dedicated to her grace the Dutchess Dowager of Rutland. Fifth edition. Lon., 1824. 8vo, cloth, pp. 354, orig. bds.

MOUBRAY (BONINGTON). A Practical Treatise on Breeding, Rearing and Fattening all Kinds of Domestic Poultry. Pheasants, Pigeons and Rabbits; also the Management of Swine, Milch Cows and Bees and instructions for the Private Brewery. Sixth edition, with considerable additions. Colored illustrations. Lon., 1830. 8vo, cloth, bds., pp. 368.

MOUBRAY (BONINGTON). A Practical Treatise on Breeding, Rearing and Fattening all Kinds of Domestic Poultry, Pheasants, Pigeons and Rabbits. Seventh edition, with considerable additions. Hand coloration of illustrations, 20 pp. of announcements of early publications. Lon., 1834. 8vo, pp. 467, orig. bds.

MOUBRAY (BONINGTON). A Practical Treatise on Breeding, Rearing and Fattening all Kinds of Domestic Poultry, Pheasants, Pigeons and Rabbits; also the Management of Swine, Milch Cows and Bees. With instructions for the private brewery, on cider, perry, and British wine making. Illustrated hand coloration illustrations, containing the announcement of early books on sports. Eighth edition with additions. London, 1842. 8vo, cloth bds., pp. 467.

MEALL (L. A.). Moubray's Treatise on Domestic and Ornamental Poultry. A practical guide to the history, breeding, rearing, feeding, fattening and general management of fowls and pigeons. New edition, revised and greatly enlarged. To which is added: The Diseases of Poultry, with Physiological Observations and Experiments, by F. R. Horner, Esq., M. D. 8 full pages, new illustrations colored by hand. Book originally opened by E. E. Richards, Aug. 25, 1909. Lon., 1854. 8vo, cloth, bds., pp. 504.

NOLAN (J. J.). Ornamental, Aquatic and Domestic Fowl and Game Birds; their importation, breeding, rearing and general management. Embellished with fifty highly finished engravings. The drawings and engravings by Mr. Wm. Oldham, with 8 page appendix describing the recently imported Cochin China or Shanghae fowl illustrated. Dublin, 1850. Limp cloth, pp. 191.

NORRIS-ELYE (L. C. R.). Brahmas and Cochins. A treatise on these popular fowl. N. d. 61 pp., boards, illustrated. The Feathered World, London.

NORTHUP (GEO. N.). Minorcas of every Comb and Color. 1907. 104 pp., boards, illustrated.

NORYS (MYRA V.). Pocket Money Poultry. Containing 20 illustrations. Washington, D. C., (n. d.). Limp cloth, pp. 167.

NOURSE (H. A.). Chicks, Hatching and Rearing. A manual of dependable instructions in incubating, brooding, feeding, housing and developing winners and layers; Fattening, killing and marketing broilers and roasting chickens. Completely illustrated. St. Paul, 1909. Bds., pp. 126.

NOURSE (H. A.). Egg Money, How to Increase It. A book of complete and reliable information. 128 pp., boards, illustrated, 1907. Webb Pub. Co., St. Paul, Minn.

PALMER (WALTER, M. P.). Poultry Management on a Farm. An account of three years' work with practical results. 94 pp., boards, illustrated, 1902. Archibald Constable & Co., Westminster. With the compliments of the author.

PAYNTER (F. G.). How to Make Poultry Pay. A practical manual. London, MCMVII. Cloth, pp. 95.

PEARSON (LEONARD, B. S. V., M. D.). Part 1, Diseases of Poultry. Illustrated. 9 pages, colored illus. With autograph on fly leaf. J. Clayton. Erh., Esq., Comp. of Jno. M. Scott. Penn., 1897. Half leather, pp. 749.

PHILO (E. R.). The Philo System of Progressive Poultry Keeping. Eighteenth edition, illustrated. Elmira, 1911. Paper covers, pp. 93.

PHILO (E. R.). The Philo System of Progressive Poultry Keeping. Sixth edition. Illustrated. Elmira, N. Y., 1908. Paper covers, pp. 64.

PHILO (E. R.). Making Poultry Pay by the Philo System. North, South, East and West. First edition, illustrated. Elmira, 1911. Paper covers, pp. 95.

PHILO (E. R.). Poultry Diseases. A handbook prepared by The Philo National Poultry Institute. Elmira, (n. d.). Cloth, pp. 92.

PIERSON (CLARA DILLINGHAM). Tales of a Poultry Farm. Illustrated. New York, (n. d.). Cloth, pp. 195.

PIPER (HUGH). Poultry. A practical guide to the choice, breeding, rearing and management of all descriptions of fowls. turkeys. guinea fowls, ducks and geese for profit and exhibition. Illustrated with eight colored plates. Second edition. London, MDCCCLXXII. Cloth, pp. 152.

PIPER (HUGH). Poultry. Their varieties, management, breeding and diseases. Comprising clear and copious directions for the construction and fitting up of poultry houses, etc. Choice of stock, obtaining eggs cheaply all the year round, the rearing, fattening and general management of fowls, turkeys, guinea fowls, ducks and geese and the treatment of their diseases. With a beautifully colored plate. London, (n. d.). Flexible limp cloth, pp. 64.

POULTRY EXPERTS of "The Smallholder." Profitable Poultry-keeping. For smallholders and others. Containing illustrations. London, paper covers, pp. 191.

POULTRY KEEPER CO. The Poultry Keeper Illustrator. Containing over one hundred illustrations. Poultry houses, incubators, brooders, nests, fences, roosts, trap devices, troughs, drinking fountains, warming appliances, coops, etc. Parkersburg, Pa., (n. d.). Paper covers, pp. 23.

POULTRY PRESS, LTD. The Poultry World Annual. Devoted to home and fancy poultry. London, 1910. Papers covers, pp. 259.

POULTRY PRESS, LTD. The Poultry World Annual. Illustrated. London, 1909. Paper covers, pp. 95.

POWELL (EDWIN C.). Making Poultry Pay. Containing illustrations. New York, 1907. Cloth, pp. 307.

PROUD (P.). Management of Incubators and Rearers. London, (n. d.). Paper covers, pp. 16.

PROUD (P.). Bantams as a Hobby. With two colored plates and upwards of thirty illustrations. Photo and autograph of P. Proud. With appendix two pages. London, (n. d.). cloth, pp. 56.

PROUD (F.). Management of Chickens from Shell to Maturity. Fifth edition. London, (n. d.). Paper covers, pp. 8.

PROUD (F.). The Game Fowl. (Old English and Modern). To which is added a reprint of The Cocker by W. Sketchley, Gent. First published in 1814. Containing supplements, The Feathered World. London, 1903. Cloth, pp. 85.

Selby, John Prideaux

PRIDEAUX (JOHN SELBY, F. R. S., F. L. S., M. W. S., etc., etc.). The Natural History of Pigeons. Illustrated by thirty full page plates in colors. Containing 82 pages of memoir of Pliny by Andrew Crichton.

PURVIS (MILLER). Poultry Breeding. A complete guide for keepers of poultry. Illustrated. Chicago, 1910. Cloth, pp. 323.

PURSELL (J. P., M. D.). Poultry Sense. A treatise on the management and care of chickens including the treatment of the more common diseases. Sellersville, Pa., 1911. Cloth, pp. 123.

PROCEEDINGS OF THE AMERICAN POULTRY ASSOCIATION—

1885. New York City, Feb. Geo. S. Joslyn, secretary, Frederic, N. Y. 32 pp., boards. H. S. Babcock, compliments of H. H. Stoddard.

1886. St. Louis, Mo., December. Walter Elliott, secretary, Shelbyville, Ind. 42 pp., boards.

1888. Indianapolis, Ind., January. Walter Elliott, secretary, Shelbyville, Ind. 120 pp., boards.

1889. Buffalo, N. Y., January. Richard Twells, secretary, Montmorenci, Ind. 98 pp., boards.

1890. New York City, February. Richard Twells, secretary, Montmorenci, Ind. 60 pp., boards.

1891. Charleston, S. C., January. Geo. E. Peer, secretary. Autograph H. S. Pelton, member A. P. A. 1874, Penn Yan, N. Y.

1892. Los Angeles, Calif., February. Geo. E. Peer, secretary, Rochester, N. Y. Autograph of H. S. Babcock.

1893. Chicago, Ill., October. Geo. O. Brown, secretary, Baltimore, Md. 80 pp., boards.

1894. Kansas City, Mo., December. Theo. Hewes, secretary, Trenton, Mo. 48 pp., boards.

1896. Washington, D. C., February. D. Lincoln Orr, secretary, Orr's Mills, N. Y. 60 pp., boards.

1897. New York, N. Y. Theo. Sternberg, secretary, Ellsworth, Kans. 62 pp., boards.

1898. Boston, Mass. Theo. Sternberg, secretary. 144 pp., boards.

1899. Toronto, Canada, January. Theo. Sternberg, secretary. 104 pp., boards.

1900. Cedar Rapids, Iowa, January. H. A. Bridge, secretary. Columbus, Ohio. 68 pp., boards.

1901. Chicago, Ill. H. A. Bridge, secretary, Columbus, Ohio. 84 pp., boards.

1901. Buffalo, N. Y. and 1902, Charleston, S. C. T. E. Orr, secretary, Beaver, Pa. 20 pp., boards.

1902. Adjourned meeting, Hagerstown, Md., and 27th annual meeting, Cleveland, Ohio. T. E. Orr, secretary, Beaver, Pa. 22 pp., boards.

1903. Adjourned meeting, Indianapolis, Ind., September, and 28th annual meeting, Rochester, N. Y., February, 1894. T. E. Orr, secretary. 100 pp., boards.

1904. Adjourned meeting, St. Louis, October, 1904, 29th annual meeting, Minneapolis, Minn., January, 1905, called meeting, Pittsburg, Pa., April, 1905; adjourned meeting, Hagerstown, Md., October, 1905, 30th annual meeting, Cincinnati, Ohio, January, 1906. T. E .Orr, secretary, Beaver, Pa. 172 pp., boards.

1907. Adjourned meeting, Auburn, N. Y., January. T. E. Orr, secretary, Beaver, Pa. 52 pp., boards.

1907. Niagara Falls, N. Y., August. Ross C. H. Hallock, secretary, St. Louis, Mo. 128 pp., boards.

1908. Niagara Falls, N. Y., August. Ross C. H. Hallock, secretary, St. Louis, Mo. 211 pp., boards.

1909. Niagara Falls, N. Y., August. Fred L. Kimmcy, secretary, Morgan Park, Ill. 182 pp., boards.

1910. St. Louis, Mo., August. S. T. Campbell, secretary, Mansfield, Ohio. 214 pp., boards.

1911. Denver, Colorado, August. S. T. Campbell, secretary, Mansfield, Ohio. 256 pp., boards.

QUISENBERRY (T. E.). The Poultryman's Guide. A book containing hundreds of practical ideas and valuable information for the beginner, the farmer, the fancier or the expert. Containing illustrations. Mountain Grove, Mo., (n. d.). Paper covers, pp. 243.

RANKIN (James). Natural and Artificial Duck Culture. Illustrated. Boston, 1889. Paper covers, pp. 96.

RANKIN (JAMES). The Incubator and Its Use. Containing illustrations. South Easton, (n. d.). Paper covers, pp. 72.

RANKIN (JAMES). Natural and Artificial Duck Culture. Fifth edition revised and enlarged. Illustrated. South Easton, Mass., 1906. Paper covers, pp. 146.

RANKIN (JAMES). Sixteen Years' Experience in Artificial Poultry Raising. Containing illustrations. Springfield, 1886. Paper cover, pp. 84.

RENWICK (E. S.). The Thermostatic Incubator, Its Construction and Management. Together with descriptions of brooders, nurseries, and the mode of raising chickens by hand. Written on the fly leaf with the compliments of the author. New York, 1883. Flexible cloth, pp. 98.

REW (R. HENRY). Royal Commission on Agriculture. England report on the poultry rearing and fattening industry of the Heathfield District of Sussex. Formerly owned by W. B. Tegetmeier, containing clippings and comments by him. London, 1895. Paper covers, pp. 32.

RELIABLE POULTRY JOURNAL. The Chick Book. Contains experience of the world's leading poultryman. 80 pp., boards, illustrated. 1905. Reliable Poultry Journal, Quincy, Ill.

RELIABLE POULTRY JOURNAL. Successful Poultry Keeping. A text book for the beginner and for all persons interested in better poultry and more of it. Illustrated in color, 176 pp., boards. 1907. R. P. J., Quincy, Ill.

R. P. J. PUB. CO. Poultry Houses and Fixtures. Up-to-date designs of practical buildings for city lot, the village acre or farm. 32 pp., boards, illustrated. 1898. R. P. J. Pub. Co., Quincy, Ill.

R. P. J. Turkeys, Ducks and Geese. Their care and management, mating, rearing, judging with explanation of the score card. 1904. 184 pp., illustrated, boards. R. P. J. Pub. Co., Quincy, Ill.

RELIABLE POULTRY JOURNAL. Artificial Incubating and Brooding. The successful hatching and rearing of poultry by modern artificial means. 96 pages, illustrated, boards. 1906. R. P. J. Pub. Co., Quincy, Ill.

RELIABLE POULTRY JOURNAL. Eggs and Egg Farms. Trustworthy information regarding the successful reproduction of eggs, construction of plans for poultry buildings. 96 pp., boards. 1907. Illustrated. R. P. J. Pub. Co., Quincy, Ill.

RELIABLE POULTRY JOURNAL CO. Breed Books. Wyandottes, by Drevensledt, Plymouth Rocks by Denny, Asiatics by Hunter and others, Leghorns various authors; Rhode Island Reds, D. E. Hale; Orpingtons, Drevensledt. 636 pp., buckram, illustrated, black and colors, 1909-11. R. P. J. Pub. Co., Quincy, Ill.

R. H. A Lover of the Sport. The Royal Pastime of cock fighting or the art of breeding, feeding, fighting and curing cocks of the game. On fly page, "Facsimile Impression of the old 1709 of which this is number 66, 100 copies only printed, May, 1899." London, printed for D. Brown, at the Black Swan without Temple-bar, and T. Ballard at the Rifling Sun in Little Britain, 1709. Orig. bds., pp. 92

RICHARDSON (H. D.). Domestic Fowl and Ornamental Poultry. Their natural history, origin and treatment in health and disease. With illustrations on wood. New York, 1852. Rebound, orig. bds., pp. 96.

RICHARDSON (H. D.). Domestic Fowl and Ornamental Poultry. Their natural history, origin and treatment in health and disease. With illustrations on wood. New York, 1863. Orig. bds., pp. 96.

RICHARDSON (H. D.). Domestic Fowl. Their natural history, breeding, rearing, feeding and general management. Illustrated. Third edition. Carefully revised by the author with two additional chapters, one upon feeding, the other offering clear and practical instructions for caponizing. Dublin, Jas. McGlashan, 21 D'Olier Street; William S. Orr & Co., 147 Strand, London, 1859. Limp cloth, pp. 108.

ROGERS (C. A.). International Association of Instructors and Investigators in Poultry Husbandry. Volume 1. Proceedings of the meetings for the years 1908, 1909, 1910. Ithaca, N. Y., 1912. Cloth, pp. 164.

ROLAND (ARTHUR). Poultry Keeping for Pleasure and Profit. New edition. London, 1892. Cloth, pp. 162.

ROBINSON (JOHN H.). Principles and Practice of Poultry Culture. Illustrated. Boston, (n. d.). Cloth, pp. 611.

ROBINSON (JOHN H.). Poultry Craft. A text book for poultry keepers. What to do. How to do it. Boston, Mass., 1907. Cloth, pp. 272.

ROBINSON (J. H.). Common Sense Poultry Doctor. Diseases symptoms and treatment for diseases of poultry. 176 pp., boards. 1907. Farm Poultry Pub. Co., Boston, Mass.

ROBINSON (J. H.). First Lessons in Poultry Keeping. First year course. 1907. 168 pp., boards. Farm Poultry Pub. Co., Boston, Mass.

ROBINSON (J. H.). First Lessons in Poultry Culture. Second year course. 1907. 160 pp., boards. Farm Pub. Co., Boston, Mass.

RUSSELL (HERBERT). The Incubator. A popular, practical treatise upon poultry hatching as a source of profit to the inexperienced. London, (n. d.). Cloth, pp. 86.

SANDO (R. B.). American Poultry Culture. A complete hand book of practical and profitable poultry keeping for the great army of beginners and small breeders. Illustrated. New York, MCMIX. Cloth, pp. 265.

SANDO (R. B.). Practical Poultry Keeping. Illustrated. N. Y., MCMXII. Cloth, pp. 171.

SAUNDERS (ALFRED). Our Domestic Birds. A practical poultry book for England and New Zealand. England, MDCCCLXXXIII. Cloth, pp. 261.

SAUNDERS (SIMON M.). Domestic Poultry. Being a practical treatise on the preferable breeds of farm yard poultry, their history and leading characteristics with complete instructions for breeding and fattening, and preparing for exhibition at poultry shows, etc. Derived from the author's experience and observation. New and enlarged edition. Very fully illustrated. New York, 1866. Limp cloth, pp. 120.

SAUNDERS (S. M.). Domestic Poultry. 1863. 120 pp., boards bound under title of "Chickens."

SCOTT (GEORGE). The "Red" Breeders' Annual 1912. Being the year book of the British Rhode Island Red Club. Third year of publication. Leeds, 1912. Cloth, pp. 101.

SCHROTER (DR. FR.). The Homeopathic Poultry Physician. Containing plain directions for the homeopathic treatment of the most common ailments of fowls, ducks, geese, turkeys and pigeons, based in the author's large experience and compiled from the most reliable sources. Translated from the German. New York, 1880. Cloth, pp. 92.

SHOEMAKER (C. C.). Standard Perfection Poultry Book. The recognized standard work on poultry, turkeys, ducks, and geese, containing a complete description of all the varieties, with instructions as to their diseases, breeding and care, incubators, brooders, etc., for the farmer, fancier or amateur. Chicago, 1902. Paper covers rebound in orig. bds., pp. 182.

SIMMINS (S.). Up-to-Date Poultry, Hatching and Rearing. Being the result of several years residence, keen observation, and daily practice in England's great poultry fattening centre, Heathfield, Sussex, wherefrom some 50 tons to 60 tons of dead fowls are sent up to the London market every week. London, (n. d.). Paper covers, pp. 95.

SIMPSON (GEO. R.). YODER (Levi). Indian Runner Ducks. Care and feeding, natural and artificial hatching and brooding. 2 vols. in one, bds., 164 pp. Illustrated, 1911. Authors also publishers.

SKETCHLEY (W.). The Cocker. Containing every information to the breeders and amateurs of that noble bird the Game Cock. To which is added, a variety of other useful information for the instruction of those who are attendants on the cock pit. Illustrated. London, 1814. Orig. bds., pp. 154.

SMITH (ALFRED). Profits from Poultry Farming. How to realize a profit of £50 per annum from 100 laying hens. Some actual experiences. Containing illustrations. London, (n. d.). Paper covers, pp. 63.

SPENCER (R. F.). Wyandottes and all About Them. Second edition. Containing illustrations. London, 1900. Paper covers rebound in orig. bds., pp. 80.

SPALDING (DR. T. B.). Standard and Commercial Poultry Culture. By artificial process, or, how to make poultry culture profitable. Chicago, 1885. Cloth, pp. 135.

STRONG (MAURICE H.). Artificial Production of Poultry. Containing full instructions, estimates of costs and profits, plans of buildings, a history of incubators from two thousand years back to the present time, and all the practical experience of its author. Illustrated. Cincinnati, 1884. Paper covers, pp. 46.

STRATHMORE (HELEN). Poultry Emancipated. First edition. Containing illustrations. Tonbridge, 1904. Paper covers, pp. 87.

STURGES (REV. T. W., M. A.). Poultry Culture for Profit. An illustrated guide to the general management of poultry, including selection, housing, hatching, rearing, feeding, marketing, etc. With frontispiece colored illustration of Buff Orpington cock. London, (n. d.) Paper cover, pp. 134.

STURGES (REV. T. W.). Poultry Culture for Profit. An illustrated guide to the general management of poultry. 1907. 1234 pp., flexible cover. M. C. Donald & Evans, London, Eng.

STURGES (REV. T. W., M. A.). The Poultry Manual. A complete guide for the breeder and exhibitor. Containing full information relative to poultry housing and general management, feeding, incubating, rearing, etc. Illustrated in color and black and white. London, 1909. Cloth, pp. 597.

SUTCLIFFE (J. H.). Profitable Poultry Farming. Describing in detail the methods that give the best results, and pointing out the mistakes to be avoided. Illustrated. London, (n. d.). Paper covers, pp. 128.

SUTCLIFFE (J. H.). Artificial Incubation and its Laws. Illustrated. Second edition. London, (n. d.). Orig. bds., pp. 144.

SUTCLIFFE (J. H.). Incubators and Their Management. Illustrated. Fifth edition, revised. London, 1905. Paper covers, pp. 128.

STODDARD (H. H.). How to Preserve Eggs. Containing autograph of H. S. Babcock. Hartford, 1885. Paper covers, pp. 48.

STODDARD (H. H.). The Book of the Bantams. A brief treatise upon the mating, rearing and management of the different varieties of bantams containing autograph of H. H. Stoddard. Hartford, 1886. Paper covers, pp. 58.

STODDARD (H. H.). The Plymouth Rocks. How to mate, rear, and judge them. Containing autograph of H. S. Babcock. Hartford, 1880. Paper covers, pp. 48.

STODDARD (H. H.). How to Win Poultry Prizes. Containing plain directions for mating, rearing and exhibiting prize show fowls. Illustrated. Containing autograph of H. S. Babcock. Hartford, 1881. Paper covers, pp. 56.

STODDARD (H. H.). The Brown Leghorns. How to mate, rear and judge them. Containing illustration and autograph of H. S. Babcock. Hartford, 1883. Paper covers, pp. 48.

STODDARD (H. H.). The White Leghorns. From the shell to the exhibition room. Containing illustration by Porter, from the Poultry World. Hartford, 1879. Paper covers. Containing autograph of H. S. Babcock.

STODDARD (H. H.). How to Feed Fowls. A treatise on the proper food for poultry, from the shell to maturity, for laying or breeding stock and for exhibition or market purposes. The kind, quality and amount fully described. Containing an illustration. Autograph of H. S. Babcock. Hartford, 1882. Paper covers, pp. 48.

STODDARD (H. H.). Poultry Diseases. Methods of preventing and curing them. Illustrated. Containing autograph of H. S. Babcock. Hartford, 1882. Paper covers, pp. 72.

STODDARD (H. H.). Poultry Architecture. How to build handsome and convenient fowl houses durably and economically. Containing autograph of J. B. Howe, Fortville, Ind. Containing illustrations. Hartford, 1879. Paper covers, pp. 59.

STODDARD (H. H.). The Book of the Dorking. A brief monograph upon the origin, varieties, breeding and management of the Dorking fowl. Containing illustrations. Hartford, 1886. Paper covers, pp. 35.

STODDARD (H. H.). How to Raise Pigeons. A brief manual upon the chief varieties of high class and toy pigeons with full instructions for their management in health and sickness. Hartford, 1886. Paper covers, pp. 48.

STODDARD (H. H.). An Egg Farm. The management of poultry in large number. With other articles. Illustrated. New York, (n. d.). Cloth, pp. 95.

STODDARD (H. H.). Domestic Water Fowl. Ducks, geese and swans. How to rear and manage them. Containing illustrations. With autograph of H. S. Babcock. Hartford, 1885. Paper covers, pp. 72.

STODDARD (H. H.). The Book of the Games. A brief treatise upon the mating, rearing and management of the different varieties of games. Containing illustrations. Hartford, 1886. Paper covers, pp. 63.

STODDARD (H. H.). Incubation. Natural and artificial, with illustrations and descriptions of incubators, modes of constructing breeders, and the best methods of rearing chickens artificially. Containing autograph of H. S. Babcock. Hartford, 1884. Paper covers, pp. 103.

STODDARD (H. H.). How to Win Poultry Prizes. Illustrated. Hartford, Conn., 1885. Paper covers rebound in orig. bds., pp. 56.

STODDARD (H. H.). The Wyandottes for the Fancier and for General Use. Containing illustrations. Hartford, 1885. Paper covers, pp. 48.

STODDARD (H. H.). The New Egg Farm. Or the management of poultry on a large scale for commercial purposes. A practical manual and reliable handbook upon producing eggs and poultry for market as a profitable business enterprise, either by tiself or connected with other branches of agriculture. Nearly 150 illustrations. Illustration and autograph of H. H. Stoddard. New York, 1901. Cloth, pp. 331.

Standards—American

HALSTED (A. M.). The Standard of Excellence. As adopted by the American Poultry Society. Being a reprint of the same as compiled and adopted by the London Poultry Club, with alterations and additions, adapting it to America. New York, original paper covers, pp. 28. Purchased of A. M. Halsted by E. E. Richards, July, 1912.

MASSACHUSETTS POULTRY ASSOCIATION. The American Standard of Excellence, as revised and amended by the Poultry Fanciers of America at their convention held in New York, May 10, 1871, and adopted by the Massachusetts Poultry Association. Containing portrait of W. H. Lockwood pasted in. Boston, 1871. Paper cover, pp. 88.

LOCKWOOD (WM. H.). The American Standard of Excellence. As revised by the Poultry Fanciers of America at their convention held in New York, Feb. and May, 1871. Giving a complete description of all the known varieties of fowls, also contains an essay on "Breeding Prize Birds for Exhibition. Presentation copy: "To O. B. Haden, Worcester, Mass., March 6, 1872, with the compliments of Wm. Lockwood, Hartford, Conn.," on fly leaf. Bound in pebbled morocco, gilt edges, hand tooled in gold sides and back. Hartford, Conn., 1871. Pp. 95.

AMERICAN POULTRY ASSOCIATION. The American Stand-of Excellence. As revised by the United Poultry Fanciers of America, convened under the auspices of the American Poultry Association at their convention held in Buffalo, N. Y., Jan. 15, 1874, giving a complete description of all the recognized varieties of fowls. Orig. paper cover, pp. 102. 1874.

AMERICAN POULTRY ASSOCIATION. The American Standard of Excellence. As revised by the United Poultry Fanciers of America, convened under the American Poultry Association at their convention held in Buffalo, N. Y., January 15, 1874, giving a complete description of all the recognized varieties of fowls. Souvenir of first edition and autograph presentation copy. Grant M. Curtis to E. E. Richards. Leather covers, pp. 102. 1874.

AMERICAN POULTRY ASSOCIATION. The American Standard of Excellence. As revised at Buffalo, N. Y., Jan. 15, 1875. Cloth cover, pp. 243. 1876.

AMERICAN POULTRY ASSOCIATION. The American Standard of Excellence. Revised at Buffalo, N. Y., Jan. 15, 1875. Limp cloth, pp. 243.

AMERICAN POULTRY ASSOCIATION. The American Standard of Excellence. As revised at convention held in Buffalo, N. Y., Jan. 15, 1875; in Chicago, Jan. 24, 1876; in Buffalo, Feb. 5, 1877; and in Portland, Maine, Feb., 1878. Contains book plate of Alexander B. King. Limp cloth, pp. 243. 1878.

AMERICAN POULTRY ASSOCIATION. The American Standard of Excellence. As revised at their annual meetings, including the changes at their annual meetings at Buffalo, Feb., 1879. Containing appendix No. 4 changes adopted Jan., 1880, at Indianapolis. Limp cloth, pp. 243. 1879.

AMERICAN POULTRY ASSOCIATION. The American Standard of Excellence, as revised at Worcester, Mass., 1883. Ninth edition. Cloth, pp. 256. 1883.

AMERICAN POULTRY ASSOCIATION. The American Standard of Perfection. Edited by Harmon S. Babcock. As adopted at Indianapolis, Ind., 1888. Obselete edition, with outline illustrations. Autograph of H. S. Babcock. Cloth covers, pp. 244. 1888.

AMERICAN POULTRY ASSOCIATION. The American Standard of Perfection. Edited by Harmon S. Babcock. As adopted at Indianapolis, Ind., 1888. Cloth covers, pp. 244. 1888.

AMERICAN POULTRY ASSOCIATION. The American Standard of Perfection. Edited by Harmon S. Babcock. As adopted at Indianapolis, Ind., 1888. Cloth, pp. 244. 1892.

AMERICAN POULTRY ASSOCIATION. The American Standard of Perfection. Illustrated. As revised at Niagara Falls, N. Y., 1909, and at its 35th annual meeting at St. Louis, Mo., 1910. Cloth, pp. 331. 1910. Contains supplementary correction list.

AMERICAN POULTRY ASSOCIATION. The American Standard of Perfection. Edited by B. N. Pierce. As adopted at Chicago, Ill., 1893. (With errata pasted in). Autograph of H. S. Babcock. Cloth, pp. 278. 1894.

AMERICAN POULTRY ASSOCIATION. The American Standard of Perfection. Edited by J. H. Drevenstedt. As adopted at Boston, Mass., 1898. Limp cloth, pp. 255. 1898.

AMERICAN POULTRY ASSOCIATION. The American Standard of Perfection...Edited by J. H. Drevenstedt. As adopted at Boston, Mass., 1898. Limp cloth, pp. 257. 1902.

AMERICAN POULTRY ASSOCIATION. The American Standard of Perfection. Illustrated. As revised at Rochester, N. Y., 1904. Cloth, pp. 299. Copyright, 1906.

AMERICAN POULTRY ASSOCIATION. The American Standard of Perfection. Illustrated. As revised at Niagar Falls, N. Y., 1909, and at its 35th annual meeting at St. Louis, Mo., 1910. Correction edition, 1910. Cloth, pp. 331. 1910.

TALLERMAN (D., F. K. J.). "Eggs." Their Collection and Sale. An address to farmers, farmer's wives, and daughters, landowners and laborers. London, 1895. Paper covers, pp. 24.

TEGETMEIER (W. B.). Prize Essay on Rearing and Fattening Market and Table Poultry. Reprinted with additions from Yorkshire Agricultural Society Reports. The Mark Lane Express, Gardner's Chronicle, The Cottage Gardner. London, 1863. 24 pp., boards. (Under title of "Chickens").

TEGETMEIER (W. B., F. Z. S.). The Standard of Excellence in exhibition poultry, authorized by the Poultry Club, to which is added the American Standard, reprinted from the original editions with additions. London, 1874. Cloth bound, pp. 113.

TEGETMEIER (W. B., F. Z. S.). Poultry for the Table and Market Versus Fancy Fowls. With an exposition of the fallacies of poultry farming. Containing book mark Walt James Lindsay. The second edition, revised and enlarged. Illustrated. London, 1893. Limp cloth, pp. 129.

TEGETMEIER (W. B., F. Z. S.). The Poultry Book. Comprising the breeding and management of profitable and ornamental poultry; to which is added "The Standard of Excellence in exhibition birds." With colored illustrations by Harrison Weir and numerous engravings on wood. London, 1873. Half leather, pp. 390.

TEGETMEIER (W. B.). The Cottager's Manual of Poultry Keeping. Being chapters from W. B. Tegetmeier's Poultry for the Table and Market Versus Fancy Fowls. London, 1893. Paper covers, pp. 46.

TEGETMEIER (W. B., F. Z. S., M. B. O. U.). On the Principal Modern Breeds of the Domestic Fowl. Containing illustrations. (From the Ibis July, 1890). Paper covers, pp. 327.

THEOBALD (FRED V., M. A., F. E. S.). The Parasitic Diseases of Poultry. With illustrations by the author. London, MDCCCXCVI. Cloth, pp. 120.

THORNE (E. C.). Revised by Jacobs (P. H.). The new and complete poultry book. A manual for the American poultry yard. 1900. 210 pp., boards. Crowell Pub. Co., Springfield, Ohio.

TILSON (IDA E.). Sewell (F. L.). The Poultry Manual. A guide to successful poultry keeping in all its branches. 144 pp., boards. 1908. Webb Pub. Co., St. Paul, Minn.

TOWNSEND (CHARLES F.). **Poultry Secrets Revealed.** Buffalo, 1911. Paper covers, pp. 87.

TURNER (DR. WM.). **Turner on Birds (1544)**...A short and succinct history of the principal birds noticed by Pliny and Aristotle. Edited with introduction, translation, notes and appendix by Evans (A. H., M. A.). Cambridge, 1903. Cloth, gilt top, pp. 221.

TREMBLEY (MR., F. R. S.). **The Art of Hatching and Bringing up Domestic Fowls,** by means of artificial heat. Being an abstract of Monsieur de Reaumir's curious work on the subject, communicated to the Royal Society. Translated from the French. Containing autograph of George Herbert Wailes. London, MDCCL. Orig. bds., pp. 61.

TWEED (ISA). **Poultry Keeping in India.** A simple and practical book on their care and treatment, their various breeds and the means of tendering them profitable. Third edition. Illustrations taken from American Poultry Journal, Reliable Poultry Journal, Harper Eng. Co. Calcutta, 1909. Cloth bound, pp. 237.

TWEED (ISA). **The Indian Handbook on Ducks, Geese, Turkeys, Guinea Fowls, Pigeons, Pea Fowls, and Rabbits.** A simple and practical book on their care and treatment, their various breeds and the means of rendering them profitable. With numerous illustrations. Calcutta, 1901. Cloth, pp. 105.

TWINNING (S. B.). **Poultry Truths.** Containing illustrations. Yardley, Penn., 1910. Paper covers, pp. 42.

TYAS (REV. ROBERT, B. A.). **Beautiful Birds Described.** With 36 illustrations by James Andrews., F. R. H. S. In three vols. London, (n. d.). Pp. 500.

VERREY (L. C.). **The Leghorn Fowl.** Containing illustrations. With written notice "Photo Eng. Co., Dear Sirs: Please make cut same size as new and oblige, Yours truly, H. H. Stoddard, Pub. Poultry World, Hartford, Ct." Page cut out and pasted in. London, 1887. Paper covers, pp. 57.

VERREY (L. C.). **Bantams.** How to breed and rear. Containing illustrations. Containing autograph of H. S. Babcock. London, E. C., 1893. Paper covers, pp. 81.

VERREY (L. C.). **The Andalusian Fowl.** Containing autograph of H. S. Babcock. Containing full page plate Andalusian cock "Ludlow." London, 1889. Paper covers, pp. 46.

VONCULIN (E. & C.) **The Art of Incubation and Brooding.** A guide to profitable poultry raising. Containing photos of E. & C. VonCulin. Illustrated. Delaware City, Del., 1894. Limp cloth, pp. 170.

WALLACE (JOSEPH). **Barred and White Plymouth Rocks.** Their history, characteristics and standard points, how to mate and rear them for exhibition and commercial purposes; with a chapter on their diseases and treatment. Containing colored and uncolored illustrations. Albany, N. Y., 1888. Paper cvoers, pp. 57.

WATSON (J. A. S., B. Sc., F. R. S. E.). **Heredity.** London, Cloth. A chapter devoted to poultry. Illustrated, pp. 90.

WATSON (GEORGE C., M. S.). **Farm Poultry.** A popular sketch of domestic fowls for the farmer and amateur. Sixth edition. Illustrated. New York, 1907. Cloth, pp. 341.

WATTS (ELIZABETH). **Poultry.** An original and practical guide to their breeding, rearing, feeding and exhibiting. With illustrations. London, (n. d.). Orig. bds., pp. 192.

WATSON (R.). **Eggs and Poultry as a Source of Wealth.** Illustrated. London, (n. d.). Paper covers, pp. 91.

WARD (C. J.). The Poulters' Guide, for Treating Diseases of Poultry. Giving cause, symptoms and remedies for their cure. Also how to caponize fowls, and feed and rear chicks hatched in an incubator. Third revised edition. Chicago, 1885. Paper covers, pp. 44.

WARREN (EDGAR). Side Line Poultry Keeping. "Two Dollars a Day from Poultry and Eggs." Revised and improved. Illustrated. Syracuse, 1909. Paper covers, pp. 94.

WARREN (EDGAR L.). 200 Eggs per Year per Hen. A practical treatise on egg making and its conditions and profits. 1902. 80 pp., boards, illus. Edgar L. Warren, Wolfeboro, N. H.

WEBB PUBLISHING CO. Poultry Houses, Coops and Equipment. A book of new plans for building practical up-to-date colony houses, continuous houses, roosting coops, brood coops, fixtures and utensils; for the farmer, the village poultry keeper and the exclusive poultry raiser. Fully illustrated. St. Paul, Minn., 1909. Paper covers, pp. 96.

WEBB PUBLISHING CO. Poultry Houses, Coops and Equipment. A book of new plans for building practical up-to-date colony houses, continuous house roosting coops, brood coops, etc. Fully illustrated. St. Paul, (n. d.) Paper covers rebound, pp. 96.

WEIR (HARRISON, F. R. H. S.). Our Poultry and all About Them. Their varieties, habits, mating, breeding, selection, and management for pleasure and profit. With chapters and notes, historical, antiquarian, traditional, proverbial, curious, instructive and interesting. In two volumes. With 576 illustrations including 36 colored plates. London, (n. d.). Cloth, gilt edges, pp. 882.

WESTON (R.). Tracts on Practical Gardening. Particularly addressed to the gentleman Farmers in Great Britain, with several useful improvements in stove and green houses. To which is added 24,674. A chronological catalogue of English authors on agriculture, botany, gardening, etc. The second edition greatly improved. London, MDCCLXXII. Orig. bds., pp. 136.

WHEELER (ARTHUR S.). Profitable Breeds of Poultry. New York, MCMVII. Cloth, pp. 134.

WHITFIELD (G. T.). The Indian Game Fowl. Containing illustration. London, (n. d.). Paper covers, pp. 38.

WILSON (F. E.). Poultry Keeping and How to Make it Pay. London, (n. d.). Cloth, pp. 126.

WILLIAMS (THOMAS B.). Farmer's Guide in the Management of Domestic Animals and the treatment of their diseases. A treatise on horses, mules, meat cattle, sheep, swine, poultry, bees, etc. Embellished with engravings. New York, 1849. Paper covers rebound, bds., pp. 100.

TEGETMEIER (W. B., F. Z. S.). The Poultry Book. Comprising the breeding and management of profitable and ornamental poultry, their qualities and characteristics. To which is added "The Standard of Excellence in Exhibition Birds." Authorized by the Poultry Club. With colored illustrations by Harrison Weir, and numerous engravings on wood. London, 1867. Limp cloth, gilt edges, pp. 356.

WINGFIELD (REV. W.). Johnson (G. W.). The Poultry Book. Comprising the characteristics, management, breeding and medical treatment of poultry. Being the results of personal observation and the practice of the best breeders including Captain W. W. Hornby, R. N., Edward Sons. Esq., Thos. Sturgeon, Esq., Chas. Punchard, Esq., Edward Hewitt, and others. With colored representations of the most celebrated prize birds drawn from life by Mr. Harrison Weir and printed in colors under his superintendence. London, 1853. Half leather, pp. 324.

WOODS (PRINCE T., M. D.). How to Raise Chicks. Including revision of facts about white diarrhoea. A practical book that tells how to select and manage breeding fowls, what you want to know about foods and feeding, etc. Illustrated. Chicago, Ill., 1912. Cloth, pp. 123.

WOODS (PRINCE T., M. D.). Open-Air Poultry Houses for all Climates. A practical book on modern common sense poultry housing for beginners and veterans in poultry keeping, what to build and how to do it, etc. Chicago, 1912. Cloth, pp. 86.

WRIGHT (LEWIS). The New Book of Poultry. With forty-five plates in colour and black and white by. J. W. Ludlow. Containing the Poultry Club Standards of Perfection for the various breeds. London, MCMV. Cloth, gilt edges, pp. 600.

WRIGHT (LEWIS). The Book of Poultry. With practical schedules for judging constructed from actual analysis of the best modern decisions. Illustrated. Popular edition. London, 1885. Cloth, pp. 591.

WRIGHT (LEWIS). The Brahma Fowl; a Monograph. Third and revised edition. Containing colored and uncolored illustrations. London, 1873. Cloth, pp. 144.

WRIGHT (LEWIS). The Practical Poultry Keeper. With eight colored plates and other illustrations. Cassell & Co., Ltd., London, Paris, New York, Toronto and Melbourne, 1905. Cloth, pp. 315.

WRIGHT (L.). The Practical Poultry Keeper. A complete and Standard Guide to the management of poultry, whether for domestic use, the market or exhibition. Illustrated. Autograph of Z. Edwards Lewis, 1875. New York, (n. d.). Cloth, pp. 243.

Poultry Publications

WESTERN POULTRY JOURNAL, Cedar Rapids, Ia. October 1888, to Sept., 1912. 23 vols., bound half leather.

POULTRY WORLD, Hartford, Conn. Vol. I, Jan., 1872, to Dec., 1872; Vol. II, Jan., 1873, to Dec., 1873; Vol. VI, Jan., 1877, to Dec., 1877; Vols. VII-VIII (chromo edition), Jan., 1878, to Dec., 1879.

POULTRY BULLETIN, New York City. Vols. I, II, April, 1870, to March, 1872; Vols. IV, V, April, 1874, to Dec., 1874.

CANADIAN POULTRY CHRONICLE, Toronto, Can. Vol. II, July, 1871, to July, 1872.

FARM POULTRY, Boston, Mass. Sept., 1888, to Sept., 1900, Vols. I to XI.

POULTRY. Peotone, Ill. Oct., 1904, to Sept., 1911. Vols. I to VII.

AMERICAN POULTRY JOURNAL. July, 1874, to Dec., 1877. Vols. I to VII.

POULTRY KEEPER, Parkersburg, Pa. March, 1897, to Feb., 1900. Vols. XI to XVI.